Agriculture Crop Production

Grade 7 & 8

Copyright @ 2021 by

Rommel Umengan Sacramento

ISBN 978-1-105-46778-3

Published by
Lulu Book Publishing

Printed in the
United States of America

Table of Contents

Chapter 1: Farm tools and equipment and their maintenance 1
1. Farm tools and Equipment 1
2. Defects and remedies 5
3. Farm tools safety 6
4. Farm equipment 9
5. Equipment Design Defects and Hazards 10
6. Farm Equipment Specification 11
7. Pre-Operational Checks for Equipment 14
8. Safety Practice During Operation of Farm Equipment 16
9. Operating A Tractor 16
10. Preventive Maintenance 17

Chapter 2: Farm estimation and basic calculation 19
Learning Outcomes 19
1. Farm inputs 20
2. Farm Labort 21
3. Labor Requirement in Planting 22
4. Estimating farm inputs and labor requirements 24
5. Estimated number of weeding operation 25
6. Perform calculation 27
7. System of measurement 34
8. Units of measurement 35
9. Metric Conversion 36
10. Fraction and decimals, percentage and ration 38

Chapter 3: Interpreting plans and drawings 43
1. Farm plans and layout 43
2. Planting system 45
3. Farm layout 49
4. Government plans 49
5. Layout plan of irrigation system 50
6. Types of irrigation systems 51
7. Essential features of a plan 54

Chapter 4. Safety measures in farm operations 57
1. Applying safety measures 58
2. Hazards, risks and exposure in the farm 59
3. Farm work involving chemicals 61
4. Personal protective equipment (PPE) 62
5. Basic first aid 63
6. Cleaning, storing and waste management 65
7. Procedure and technique in storing materials and chemicals 67
8. Government requirement regarding farm waste disposal 71
9. Water management system 72

Chapter 1
Farm Tools and Equipment

❖ **Learning Outcomes**

1. *Select and use farm tools* [1]
    1.1. Identify farm tools according to use
    1.2. Check farm tools for faults
    1.3. Use appropriate tools for the job requirement according to the manufacturer's specifications and instructions

2. Select farm equipment [1]
    2.1. Identify appropriate farm equipment
    2.2. Follow the guidelines in the instructional manual of farm equipment
    2.3. Conduct pre-operation check-up in line with the manufacturer's manual
    2.4. Identify faults in farm equipment and facilities
    2.5. Use farm equipment according to their function

Farm Tools and Equipment

Agricultural operations heavily rely on farm tools, implements, and equipment. Their availability makes the task run far more efficiently and effectively. However, even if one has the most advanced tools and implements, they are useless if one does not know how to utilize them. Before using agricultural tools, implements, or equipment, it is essential to understand how to use them.[2]

1. Farm Tools

Bolo is used for cutting tall grasses and weeds and chopping branches of a tree.[3] [40]

Source: amazon.ca

Crowbar is used for digging big holes and fir digging out big stones and stumps.[3]

Pick-mattock is used for digging canals, breaking hard topsoil, and digging up stones and tree stumps

It is used for breaking topsoil and pulverizing soil. [3] [39]

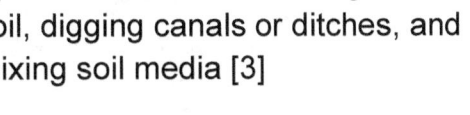

Spade is used for removing trash or soil, digging canals or ditches, and mixing soil media [3]

The shovel is used in removing trash digging loose soil, moving soil from one place to another, and mixing soil media

Rake is used for cleaning the ground and leveling topsoil

Spading Fork is used for loosening the soil, digging outcrops, and turning over the materials in a compost heap

Hand Trowel is used to loosen the soil around the growing plants and put little manure fertilizer in the soil. [3]

Hand cultivator is used for cultivating the garden plot y loosening the soil and removing weeds around the plant

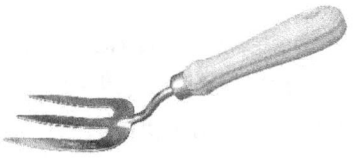
Hand fork is used for inter-row cultivation

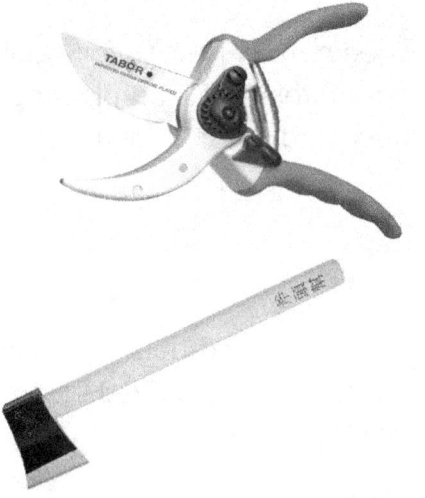
Pruning shears are for cutting branches of planting materials and unnecessary branches of plants

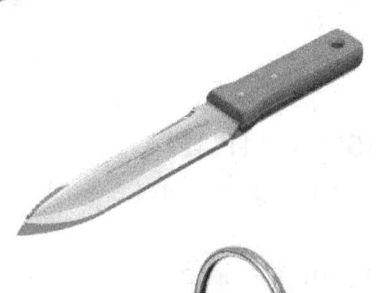
Axe is for cutting bigger size post

The knife is for cutting planting materials and for performing other operation in horticulture

Sprinklers are used for watering seedlings and young plants

Water pails are used for hauling water, manure, and fertilizers

Sprayers are a hand-held agriculture tool with a variously curved blade typically used for cutting weeds

The wheelbarrow is used for hauling trash, manures, fertilizer, planting materials, and other equipment

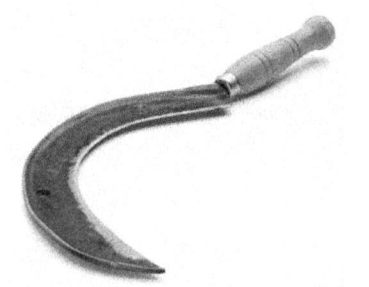
The sickle is a hand-held agricultural tool with a variously curved blade typically[4] [38] [43] [44]

FARM TOOL DEFECTS AND REMEDY[5] [6] [41] [42]
1. If the handle is broken, remove the damaged handle and replaced a new one.
2. Blunt tools can prevent from getting the desired result. A regular sharpening routine is needed.
3. Most hand tools are susceptible to rust. To avoid rusting, apply a light coat of oil over the entire metal surface of the tool. Wipe off the excess oil.
4. Lubricate the moving parts or joints to prevent wear and misalignment.

Farm Tools Maintenance
The following maintenance techniques or precautions should be followed to extend the life and effectiveness of farm tools.

1. After each use, the tools should be washed or cleaned.

2. To reduce friction, lubricate moveable parts using oil, grease, or lubricant.
3. Sharpen blunt-edged or bladed tools regularly.
4. Cutlass, for example
5. If metal parts are to be preserved for a long time, they should be painted, oiled, or greased to avoid rusting.
6. Tools should be kept in a cool, dry location.
7. Farm tool parts that are worn out or damaged should be replaced.
8. Check and tighten loose nuts and bolts every day or regularly.
9. To avoid cracking, keep instruments with wooden parts out of the sun.
10. To minimize rusting, keep metallic tools away from rain and damp areas.
11. Keep wooden-handled equipment away from termites and fire.
12. Handle tools with care and only use them for the tasks at hand on the farm.[7] [46] [50]

GENERAL FARM TOOL SAFETY

Keeping tools in good operating order is half the equation for safety. The other half is the person's ability and awareness when using the tools.

Good Working Tools = Safety + Capable and Conscious Worker

Some farm mishaps are caused by tool failure; however, most agricultural accidents are caused by weary, stressed, rushed, distracted, or inept workers.

There are basic standards for tool safety, in addition to the specific safe handling regulations for each type of farm tool:

- Read and follow the operator's safety instructions for each piece of farm tool.
- Prepare for safety by wearing suitable clothing, getting adequate rest, avoiding alcohol, and ensuring that all workers have been educated and can use the agricultural tool safely.

- When using the tool, keep all guards, shields, and access doors in place.
- Be mindful of what you're doing and where you're going.
- Adapt the tool to the working environment.
- Children and others should be kept away from the working area.
- Take pauses from work as needed.[8] [45] [47]

1. **Farm Implements**

These are attachments that are dragged by working animals or mounted on machines (hand tractor, tractor) that are commonly used for land preparation. These are typically made of a unique metal.

Plow

These are farm implements that are either pulled by a working animal or a tractor in horticultural operations. The plow is used for tilling huge regions, making furrows, and cultivating between rows. Working animals pull plows that are composed of a combination of metal and wood or pure metal. They are used to till regions that are shallower in-depth than disc plows pulled by tractors.[9]

Native Plow

Disc Plow

Harrow

The native wooden harrow is composed of wood with a metal tooth. It is pushed by a carabao, whereas the metal disc harrow is attached to a tractor. Harrows are used to pulverize and till the soil.

Native Wooden Harrow

Disc Harrow

Rotavator

The rotavator is an implement attached to a tractor used to till and pulverize the soil.

Cultivator

This is used on a field to perform supplementary tillage. Most of the time, this machine has a rotational motion.

Seed Drill

The seed drill is an attachment used for sowing seeds evenly spaced and at the proper depth. [47] [48] [49]

2. Farm Equipment

These are the machinery used in crop production. They are used in land preparation and transporting farm inputs and products—this equipment a highly-skilled operator to use.[10]

Hand Tractor

A hand tractor is used to prepare an extensive area of land by pulling a plow and harrow.

Four-Wheel Tractor

A four-wheel tractor pulls a disc plow or a disc harrow to prepare a larger area of ground.

Water Pumps

They are used to draw irrigation water from a source.

Equipment Design Defects and Hazards

Farm machinery is widely recognized as being extremely hazardous to use on the job. Many dangers have seemed to be caused by defective designs.

Farmers, for example, are frequently injured or killed when machinery without guards, shields, or warnings is sold. The following are some examples:

a. Pinch points are points on the machine where two objects make contact, and at least one of them moves in a circle.
b. Wrap points are any exposed, rotating machine component connected to a piece of equipment at one end. These can grab clothing and limbs and pull them into the machine.
c. Crush points occur when two or more things collide, or when a moving object collides with a stationary object, as when a moving tractor collides with a grain bin. Between these objects, workers can be crushed.
d. Free-wheeling points—After being turned off, some farm equipment elements, such as rotary motor blades and baler flywheels, continue to rotate.
e. Spring points—Springs carry energy that can cause injury when released.
f. Chains—Chains are frequently used to secure and pull objects on farms. If the linkages are weak or loose, they can snap and cause a collision.
g. Pull-in points—If too many crops are fed into rollers and other input sections of machinery too quickly, they can clog. Workers who

attempt to clear a jam while moving parts are at risk of being sucked into the equipment.

h. Shear points—When the edges of two sharp objects, such as those found on harvesting machines, move swiftly toward each other, workers might be wounded or maimed.

i. Hydraulic lifts and hoses—Hydraulic lifts that lift and support heavy things and help in steering and braking equipment might fail, crushing the vehicle. Hydraulic hoses that are leaking can release poisonous fluids that can cause a fire. They can also fire high-pressure oil blasts that can cause eye and skin burns.[11]

Farm Equipment Specification

Hand Tractor

This walking tractor/power tiller is equipped with a 12hp/14hp Kubota diesel engine, one-cylinder, water-cooling type with radiator, two-wheel drive with wise clock system; There are a total of eight gears: six forward and two reverse. It can do various farm tasks when equipped with various equipment, including tilling, rotavating, irrigating, planting, and hauling. It's a multi-purpose tractor that'll come in handy for farmers.[12]

SPECIFICATIONS FOR 12hp/14hp Kubota type walking tractor/power tiller			
Engine	Engine model	kW/rpm	8.82/2200
	No. of cylinders	Pcs	1 Cylinder, 4 Stroke
	Displacement	L	0.651
	Wheel Tread	mm	680--740

Main performance parameters	Tire Model	°	6.00-12
	Belts Model		B1850
	Gears		6 Forward, 2 Reverse
Gearbox system	Clutch		3 Friction Plate
	Max. Speed	km/h	17.08

4-Wheel Drive Tractor [13]

5045D MFWD (John Deree)

Engine		Wheels and Tire	
Type	45 HP, 2300 RPM, 3 Cylinder, Direct injection, In-line pump, Liquid-cooled with overflow reservoir, Water Separator	Front Rear	8.0*18, 6 PR 14.9*28, 8 PR
Air Filter	Dry type, dual element	Fuel Tank Capacity	
Transmission		60 Liters	
Clutch	Dry type dual clutch	Electrical System	
Gear Box	8 Forward + 4 Reverse speeds collar shift	88 Ah, 12 volt battery 40 Amp, alternator 12 volt, 2.5 kW starter motor, 7 pin connector	
Speeds	(3.1-32.3) Forward kmph (4.1-14.4) Reverse kmph	**DIMENSIONS AND WEIGHT**	

Brakes		Total weight	1,975 kg
Self-adjusting, self-equalizing, hydraulically actuated oil-immersed disc brakes		Wheelbase	1,950 mm
Hydraulics		Overall length	3,350 mm
Lifting capacity	1400 kgf at lower link ends	Overall width	1,765 mm
3 Point linkage	Category II	Ground clearance	460 mm
Hydraulic flow pump	46 LPM automatic depth and draft control Single SCV with couplers	ROPS	Deluxe seat with ROPS and set belt
Steering			
Power steering type			
Front Axle			
Paddy sealed MFWD Axle			
Power Take-Off			
Type	Dual PTO		
RPM	540 @ 1,705 & 2,200 ERPM		

Farm Tractor Parts and Functions

1. **Drawbar** - The drawbar is a crucial component because it is responsible for conveying huge loads. The tractor's principal function is the drawbar. This tool includes a feature that allows you to connect the tractor to the thing you want to drag, such as a straw or a trailer. This drawbar can stretch elastic in all four corners of the circle.

2. **Wheels & Tracks** - The wheels and tracks on a farm tractor are built and utilized to balance the tractor's weight so that the tract does not sink into moist soil. The tractor's wheels are larger than the front wheels, responsible for carrying the tractor's and load's weight.[14]

3. **Hitch** - The purpose of this section is to make it easier to transfer, remove, and lessen heavy loads carried by the tractor. This part is one of the farm tractor parts and their functions that is quite vital, especially when you need to convey an object from one location to another. If a hitch occurs, simply unload the thing and push it to a different site.[15]

4. **Transmission** - This element allows the tractor parts to perform more efficiently while also being more robust. The transmissions on this sidewalk were formerly shaped or operated manually, but they have recently evolved into an automatic transmission, which is more effective.

5. **Control** - Control is the next set of farm tractor parts and their functions. When the tractor is carrying a large load, this part assists the engine in driving the tractor. This control includes brakes, making it easier for the tractor to stop when it's moving down or up a steep slope. As a result, the tractor will not readily tip over, and the power will be consistent.

6. **Power take-off (PTO)** - This phase is typically accomplished by wrapping a rubber belt around the tractors powered by a pedal on the back tire. Almost all modern tractors have a power or power take-off engine, often known as a PTO, which disconnects the power. Typically, this equipment may be simply connected to the farm tractor via the wheels and gears. When the part is attached, the PTO will be able to deliver a tremendous amount of power to the tractor and the broiler engine, causing the wheels to move automatically. As a result, a tractor must have a powerful motor and a low speed so that it may be employed on agricultural land. [16]

Pre-Operational Checks for Equipment

Before heading out to the field, perform a pre-operational check on your tractors. By thoroughly inspecting your tractor before operating it, you may avoid costly repairs, downtime, aggravation, and accidents. Before starting the tractor, go over the following items on the checklist:

1. **Fuel level** - You should always make sure you have enough fuel in the tank.

2. **Check your battery** - to ensure that the terminals are not rusted.

3. **Check your tires** - In addition to checking the air pressure, ensure sure the lug nuts are snug and inspect the tire condition (e.g., tread). If you discover that your tire pressure is low, search for air leakage around the valve stem.

4. **Loose or defective parts** - Take the time to thoroughly inspect the tractor for any loose or faulty parts, such as a frayed or worn fan belt. Before traveling out to the field, replace, tighten, or make any necessary repairs.

5. **Fluid leaks** – Check the ground beneath the tractor for any fluid leaks. Check the levels of coolant, engine oil, and hydraulic oil as well. If you run out of these fluids, you could do significant harm to your tractor.

6. **Operator's platform area** - You may spend a significant amount of time on the operator platform, so make sure you can securely get on and off the tractor. Examine the area around the seat for any debris or tools that could cause you to trip. It would be best to have a ROPS on your tractor; therefore, always make sure your seat belt is functional and buckled.

7. **Fire extinguisher** - Make sure your fire extinguisher is charged.

8. **Lighting/flashers** - Check headlights and warning lights/flashers to ensure that all lights are operational and replace bulbs as needed.

9. **Visibility from the driver's seat** - Clean any dirty cab windows to provide you with the finest visibility from the driver's seat.[17] [18]

Safety Practice During Operation of Farm Equipment [19] [20] [21]

1. Always wear your seat belt. Most tractors have a rollover protection feature that protects you if the tractor rolls over.
2. Never get too close to a spinning PTO! The PTO rotates with enormous force, as does the driveshaft it's connected. Keep garments and body parts away from the rotating shaft to avoid death and mutilation!
3. Do not rest your arms or hands-on any hydraulic part's joints!
4. When the tractor is running, never put your hands inside or around active implements.
5. When using PTO-driven implements, turn off the tractor if any blockage develops.
6. Do not drive a tractor down a steep incline! Be mindful of the dangers of a rollover.
7. When lifting large loads in the bucket, always move cautiously! Driving a tractor with a heavy load greatly affects the tractor's balance and stability. Heavyweights increase tipping and rolling.

Operating A Tractor [22] [23] [24]

a. When Operating a Tractor

- Drive at a slow enough speed to maintain control over unanticipated circumstances.
- Reduce your speed before turning or applying your brakes.
- Avoid ditches, logs, rocks, depressions, and embankments.
- To increase safety on steep slopes without a trailing implement, reverse up.
- Always engage the clutch gently, especially when traveling uphill or pulling.
- On sloping terrain, use as broad a wheel track as feasible.
- Descend steep slopes slowly in low gear, utilizing the motor as a brake.
- Never mount or dismount a tractor that is moving.
- Before dismounting, make sure the parking brake is engaged and working correctly. Take brief breaks regularly if you work long hours.

b. When Towing an Implement

- Attachments should be installed following the manufacturer's instructions.
- Attach implements to the drawbar or the mounting points given by the manufacturer at all times.
- Unless otherwise specified by the manufacturer, never alter, adjust, or elevate the height of the drawbar.
- Check safety pins on towed left-wing regularly to ensure they are not worn and ensure all guards on towed implements are in place before work.
- Never hitch over the rear axle's centerline, around the axle housing, or to the top link pin, and never modify or operate on implements while they are in motion.
- Attach implements only if the PTO shaft is protected.
- Always lower the three-point linkage and towed implement when parked. [25] [26] [27]

c. **To avoid injury**
- The tractor seat can be adjusted for back support and comfort.
- When buying a tractor, be sure the seating is both secure and comfortable.
- Examine the seat height, seat depth, backrest height and angle, fore-and-aft movement, seat tilt, firm padding, partial pivoting (if you are required to look behind you for extended periods), and vibration-absorbing suspension.
    - Every hour or so, dismount and spend 5 or 10 minutes doing something physical.
    - Ensure your next tractor has low stairs, handgrips, enough entryway and cab room, and a safe mounting platform.
- Dismount by climbing down, not jumping down, and making use of every foothold and handhold available.

Preventive Maintenance [28] [29]

Preventative maintenance reduces stress more than reactive maintenance, resulting in a more efficient operation contributing to profitability. The procedures for preventive care are as follows.

1. **Implement a Maintenance Program**
   The manufacturer's recommended maintenance regimens are a good place to start with this method. It might be as simple as setting aside time to replace blades and bits, test safety components, and repair any damaged wires. Manufacturers who do not have the time to maintain power tools in-house can transfer them to authorized service centers for routine maintenance. [30] [31]

2. **Proper training on how to use the equipment**
   Workers who operate farm equipment must be well trained. This training should cover best practices for whatever job they are in and proper inspection, maintenance, and cleaning processes. Should specific equipment be used just in certain situations? Is it possible to service some tools when they are turned on or do they have to be turned off all the time? Is there a standard procedure for testing operator controls? These are the types of questions that appropriately qualified employees should be able to answer. [32] [33]

3. **Keep Them Lubricated**
   Another technique to extend the life of chains and cables is to lubricate them according to the manufacturer's instructions. Friction on these moving parts might cause your machine tool to break down if they are not adequately lubricated.

4. **Keep Them Sharpened**
   If your machine tool includes components for cutting, slicing, or sharpening, you must examine the sharpness of your equipment regularly. Wear and tear on these sharpening parts can not only require your machine to work harder, causing additional wear and tear.

5. **Keep Them Clean**
   If your machine tool is not clean, problems may develop in both the short and long term. Machines that become clogged with potentially flammable or dangerous materials can endanger both your employees and your machinery in the near run. [34] [35] [36] [37] [38]

Chapter 2
Farm Estimation and Basic Calculation

Learning Outcomes

1. Perform estimation

    1.1. Identify job requirement from written and oral communications

    1.2. Estimate the quantity of materials and resources required to complete a work/ task

    1.3. Estimate time needed to complete a work/activity

    1.4. Calculate the duration of work completion

    1.5. Follow procedures in reporting to appropriate persons the estimate of materials and resources [1] [3] [4]

2. Perform basic workplace calculation

    2.1. Identify calculations to be done according to the job requirement

    2.2. Determine correct methods of calculation

    2.3. Ascertain systems and units of measurement to be followed

    2.4. Perform calculations needed to complete a task using the four basic mathematical operations

    2.5. Employ different techniques in checking the accuracy of the result

    2.6. Use appropriate operations to comply with the job requirement [1] [2] [6]

The input type is endless, but we divide all inputs into two categories; inputs for consumables and inputs for capital. Consumable inputs for smallholder farmers – seed, fertilizers, insecticides, etc., are commonly used in the agricultural sector. Usually, consumable inputs are natural materials that are "consumed" by crops. Capital inputs, however, are often larger investment materials like tractors and trellising materials. [5] [7] [10]

Consumable Inputs

Seeds

Source: libraries.indiana.edu

Seedlings

Fertilizer

Source: thespruce.com

Capital Inputs

Tractors

Harrow

Plows

Insecticides

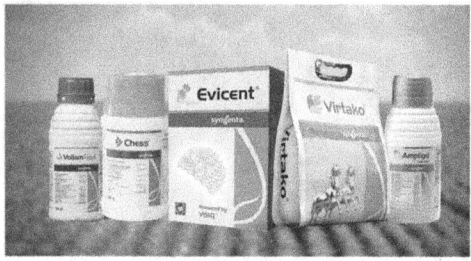

Source: Syngenta.co.in

Irrigation System

Land Preparation

Plowing using tractor

Source: nelsontractor.com

Clearing the land using hoe

Plowing with animal

Source: steemit.com

Harrowing using hand tractor

Source: reliefweb.int

Source: flickr.com

Mulching　　　　　　　　Making furrow using tractor

Source: Caption Tractors

Source: svz.com

Labor Requirement in Planting [8] [9] [11] [12]
Production of Seedlings　　　　　　Transplanting

Photo by Zoe Schaeffer on Unsplash

Image by Alex Raths

Labor Requirement for Plant Care
Fertilizer Application　　　　　　　　Pest Control

Source: knowledgebank.irri.org

Source: worldatlas.com

Weeding

Irrigation

Source: tygershark.nyc

Source: agriculture.go.ug

Harvesting

Source: analyticsindiamag.com

Estimating Farm Inputs and Labor Requirements

Irrigation expenses are the function of water per volume, the number of volumes per day, and the total number of days to irrigated from planting to the last harvest. Expressed as

$$\text{Irrigation Expenses} = \frac{\text{Price of Water}}{\text{Volume}} \times \frac{\text{No of Volumes}}{\text{Day}} \times \text{Total No days}$$

[13] [14]

Estimated Worker Hired to perform Irrigation from Planting to the Last Harvest

$$\text{Estimated Workers} = \frac{\text{Worker}}{\text{Sq. of area}} \times \text{Total Irrigated Area}$$

Estimated number of Days for Spraying Insecticide (per worker)

$$\text{Estimated no of days} = \frac{\text{No of days}}{\text{Sq. of area}} \times \text{Total Land Area}$$

[16] [17] [19]

Estimated number of Days for Spraying Insecticide (in one day)

$$\text{Estimated workers} = \frac{\text{No of worker}}{\text{Sq. of area}} \times \text{Total Land Area}$$

Estimated Cost of Insecticide use for Spraying

$$\text{Estimated Cost} = \frac{\text{Price}}{\text{Insecticide}} \times \frac{\text{No of insecticide}}{\text{Sq of area}} \times \text{Total land area}$$

Workers' Salary during insecticide spraying

$$\text{Worker' Salary} = \frac{\text{Salary}}{\text{Day}} \times \text{Total no of days}$$

[20] [22]

Estimated number of workers needed in weeding [15] [17]

$$\text{Estimated workers} = \frac{\text{No of worker}}{\text{Sq area}} \times \text{Total land area}$$

Workers Salary during weeding

$$\text{Worker' Salary} = \frac{\text{Salary}}{\text{Day}} \times \text{Total no of days}$$

Estimated number of workers employ during harvesting

$$\text{Workers employed} = \frac{\text{No of workers}}{\text{Square area}} \times \text{Total land harvesting area}$$

[21] [23]

Sample Computation

Total land area	20,000 sq meter
Amount of fertilizer	20/kilo
Number of days consumed in planting the area	2 days
Number of workers planted area	3 workers
Amount if salary paid in planting the area	300/day
Number of workers who fertilized the area from planting up to the date this visitation	150 kilos
Amount of Salary paid in fertilizer application from plantings the date of this visit	300/day
Quantity of fertilizer to be used after the survey until final harvesting	250 kilos
Number of workers required to perform fertilization after this visitation until final harvesting	2 workers
Amount of salary paid in fertilizer application from planting to this visit	300/day

[24] [25] [29]

a. The total amount of salary paid in planting
   The total amount of salary = (no of days) (no of workers) (amount of salary)
   Total amount of salary pain in plating the area = (2) (3) (300) = 1,800.00

b. The total amount of fertilizer consumed from planting up to date of the visit
   The total amount of fertilizer = (amount of fertilizer per kilo) (no of kilos)
   Total amount of fertilizer = (20) (150) = 3,000.00

c. The total amount of salary paid in fertilizing the area from planting up to date of the visit
   The total amount of salary = (no of days) (no of workers) (amount of salary)
   Total amount of salary paid in fertilizing the area = (2)(2)(300) = 1,200.00

d. The total amount of fertilizer consumed after the visit until the final harvesting
The total amount of fertilizer = (amount of fertilizer per kilo)(amount of salary)
Total amount of fertilizer = (20)(250) = 5,000.00

e. The total amount of salary paid in fertilizing the area after the visit until final harvesting
The total amount of salary = (no of days)(no of workers)(amount of salary)
Total amount of salary paid in fertilizing the area = (2)(2)(300) = 1,200.00 [26] [27] [28] [30]

Perform Calculation

It is essential to be able to measure and calculate surface areas. It might be necessary to calculate, for example, the surface area of the cross-section of a canal or the surface area of a farm.

The calculation of some of the most common surface areas will be discussed.

The most common surface areas

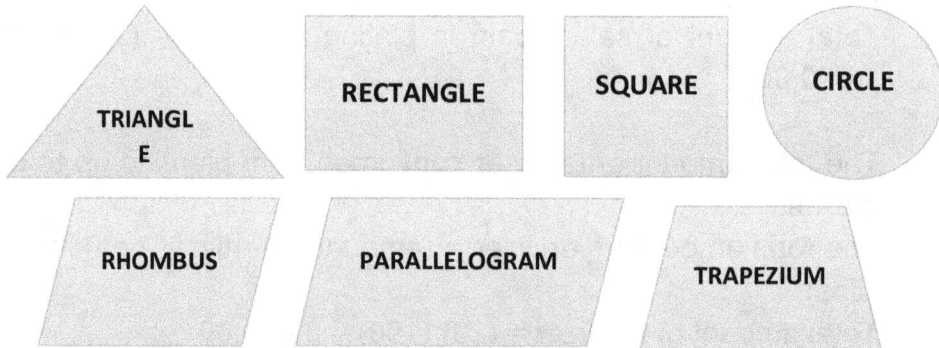

The height (h) of a triangle, a rhombus, a parallelogram or a trapezium, is the distance from a top corner to the opposite side called base (b). The height is always perpendicular to the base; in other words, the height makes a right angle with the base. An example of a right angle is the corner of this page.

In the case of a square or a rectangle, the expression length (l) is commonly used instead of base and width (w) instead of height. In the case of a circle the expression diameter (d) is used. [31] [32]

The height (h), base (b), width (w), length (l) and diameter (d) of the most common surface areas

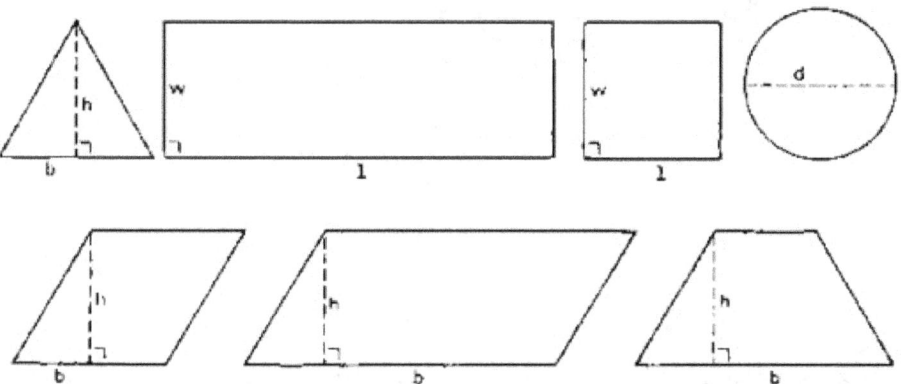

Triangle

The surface area or surface (A) of a triangle is calculated by the formula:

A (triangle) = 0.5 x base x height   (1)

Triangle can have many shapes but the same formulas is used for all of them [33] [34] [35]

Some examples of triangles

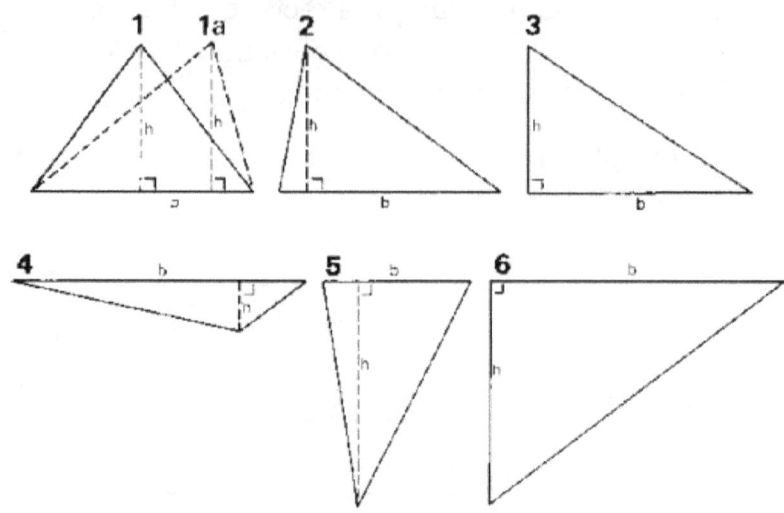

Question 1. Calculate the surface area of the triangle no 1, no. 1a and no. 2?

Given
Triangle no 1 and no 1a

Base = 3 cm
Height = 2 cm

Answer:
A = 0.5 x base x height
= 0.5 x 3 cm x 2 cm
= 3 cm²

Triangle no. 2

Base = 3 cm
Height = 2 cm

A = 0.5 x base x height
= 0.5 x 3cm x 2 cm
= 3 cm²

[3] [9]

Note: Triangle 1 no 1, no 1a and 2 have the same surface; although the shapes of the triangle are different but the base and height are in all three cases are the same, so the surface area are all the same and expressed in square centimeters (written as cm²).

Question 2. Calculate the surface areas of the triangles nos. 3, 4, 5 and 6.

Given
Triangle no 3
Base = 4 cm

Surface Area
A = 0.5 v base x height
= 0.5 x 4 cm x 3 cm

Height = 3 cm   = 6 cm²

Triangle no. 4   A = 0.5 x base x height
Base = 5 cm   = 0.5 x 5 cm x 2 cm
Height = 2 cm   = 5 cm²

Triangle no. 5   A = 0.5 x base x height
Base = 3 cm   = o.5 x 3 cm x 4 cm
Height = 4cm   = 6 cm²

Triangle no. 6   A = 0.5 x base x height
Base = 5 cm   = 0.5 x 5 cm x 4 cm
Height = 4cm   = 10 cm² [40] [41]

SQUARES AND RECTANGLES

The surface area or surface (A) of a square or a rectangle is calculated by the formula:

    A = (square or rectangle) = length x width = l x w   (2)

In a square, the lengths of all four sides are equal, and all four angles are right angles.

In a rectangle, the lengths of the opposite sides are equal, and all four angles are right angles. [2] [3] [4]

A square and a rectangle

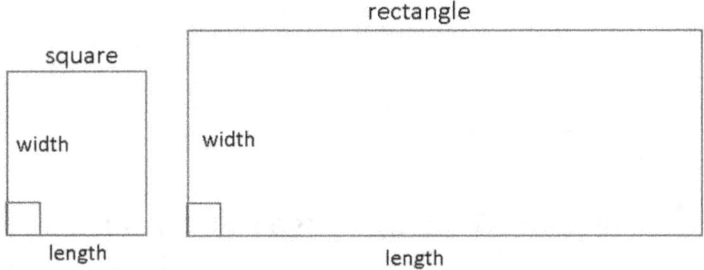

Note : In a square, the length and width are equal, whereas in a rectangle, the length and the width are not equal.

Question 3. Calculate the surface areas of the rectangle and of the square.

Given	Surface Area
Square Length = 3 cm Width = 3 cm	A= length x width = 3 cm x 3 cm = 9 cm²
Rectangle Length = 6cm Width = 4 cm	Area = length x width = 6 cm x 4 cm = 24 cm²

[5] [6]

In relation to agriculture, oftentimes you will come across the expression of hectare (ha), which is a surface area unit. By definition, 1 hectare is equal to 10,000 m². For example, a farm with a length of 100 m and a width of 100 m has a surface area of 100 m x 100 m = 10,000 m² = 1 ha.

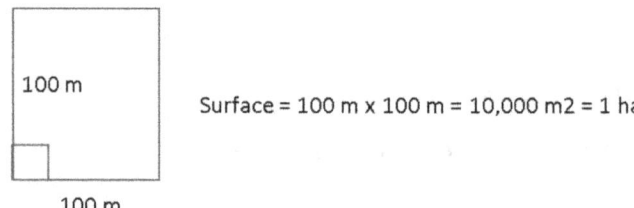

RHOMBUSES AND PARALLELOGRAMS

The surface area (A) of a rhombus or a parallelogram is calculated by the formula:

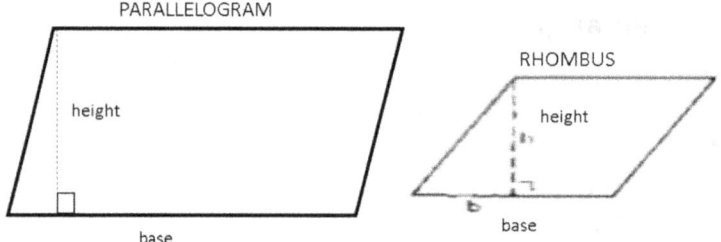

A (rhombus or parallelogram) = base x height = b x h ... (3)

In a rhombus, the lengths of all four sides are equal; none of the angles are right angles; opposite sides run parallel.

In a parallelogram, the lengths of the opposite sides are equal; none of the angles are right angles; opposite sides run parallel. [7] [8]

Question 4: Calculate the surface areas of the rhombus and the parallelogram.

Given	Surface Area
Rhombus Base = 4 cm Height = 3 cm	A = base x height = 4 cm x 3 cm = 12 cm²
Parallelogram Base = 4.5 cm Height = 3.5 cm	A = base x height = 4.5 cm x 3.5 cm =15.75 cm²

TRAPEZIUMS

The surface area of a trapezium is calculated by :

A (trapezium) = 0.5 (base + top) x height

= 0.5 (b + a) x h   (4)

The top (a) is the opposite and parallel to the base (b). In a trapezium, the base and top run parallel.

Examples of trapeziums

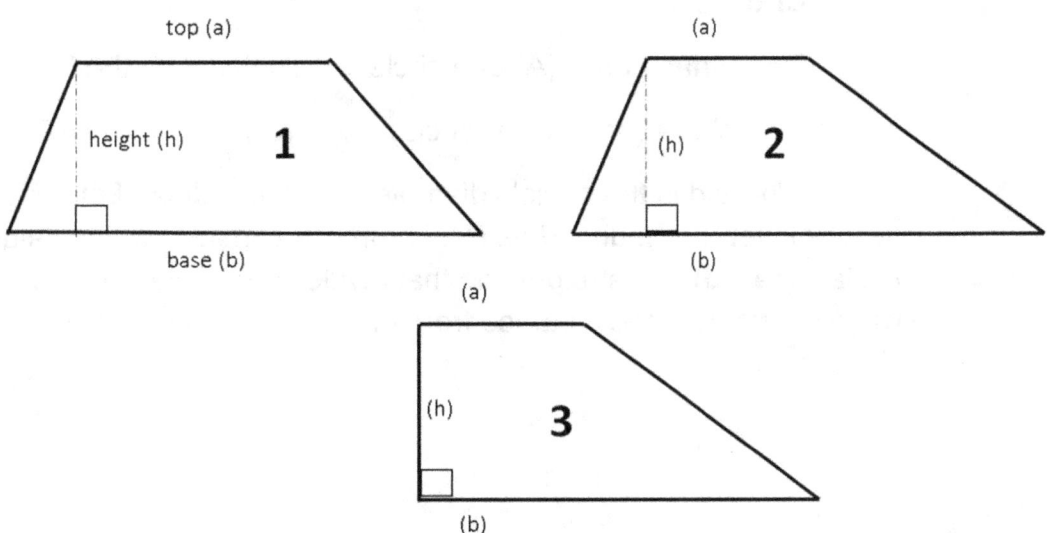

Calculate the surface area of trapezium no. 1, no. 2 and no. 3.

Given
Trapezium no 1.
Base = 5 cm
Top = 3 cm
Height = 4cm

Surface Area
Area = 0.5 x (base + top) x height
= 0.5 x (5cm + 3cm) x 4cm
= 16 cm²

Trapezium no. 2
Base = 5 cm
Top = 2 cm
Height = 4 cm

Area = 0.5 x (base + top) x height
= 0.5 x (5 cm + 2cm) x 4cm
= 14 cm²

Trapezium no. 3
Base = 5 cm
Top = 3 cm
Height = 3 cm

Area = 0.5 x (base + top) x height
= 0.5 x (5 cm + 3 cm) x 3 cm
= 12 cm²

[8] [12]

Another way to calculate the surface area of a trapezium is to split trapezium into a rectangle and two triangles. Determine the surface areas of the rectangle and the two triangles separately.

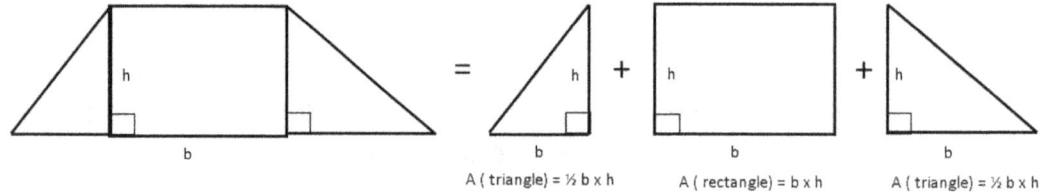

A (triangle) = ½ b x h   A (rectangle) = b x h   A (triangle) = ½ b x h

CIRCLES

The surface area (A) of a circle is calculated by the formula:

A (circle) = (πr²) or A (circle) = ¼ (πd²)   ... (5)

Where d is the circle's diameter, r is the radius of the circle, and π (Greek letter, pronounced as Pi). Π (pi) is a constant with a value of 3.14. A **diameter (d)** is a straight line that divides the circle into two equal parts while radius (r) is the distance from the center to any point on the circle's circumference.

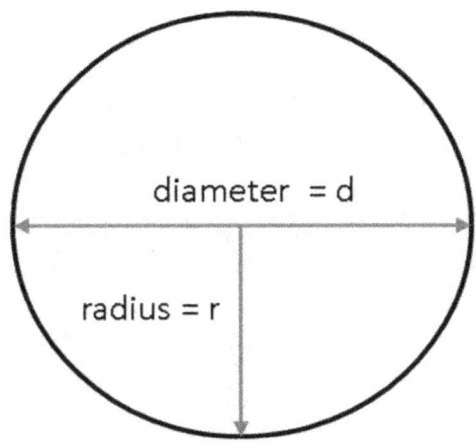

Calculate the surface area of a circle.

Given	Surface Area
Circle D = 5 cm	Area = ¼ ($\pi \times d^2$) = ¼ (3.14 x 5 cm x 5 cm) = 19.625 cm²
Circle R= 3.5 cm	Area = $\pi \times r^2$ = 3.14 x 3.5 cm x 3.5 cm = 38.465 cm²

[24] [27]

SYSTEM OF MEASUREMENT

MKS System – is a physical measurement system that uses meter, kilometer, and second as base units. It forms the base of the international System of Units

CGS System – is a variant of the metric system based on centimeter as the unit for length, the gram for the unit of mass, and the second as the unit of time.

FPS System – is a system of unit build on three fundamental units; the foot for length, the pound for either mass or force, and the second for time.

SI System – is a series of units that are accepted and used throughout the scientific world. There are seven basic units in this system; the meter (m), kilogram (kg), the second (s), the kelvin (K), the ampere (A), the mole (mol)n, and the candela (cd).

Units of Measurements

All measuring systems have basic units for length, mass, capacity (volume), and temperature. The following are the basic metric units

Basic Metric Units

Type of Measurement	Unit	Symbol
Length (distance)	meter	M
Mass (weight)	gram	G
Capacity (volume)	liter	L
temperature	Degrees celsius	°C

Units of length

Unit	Abbreviation	Length (Distance)
kilometer	km	1000 meters
hectometer	hm	100 meters
decameter	dam	10 meters
meter	m	1 meter
decimeter	dm	0.1 meter
centimeter	cm	0.01 meter
millimeter	mm	0.001 meter

Unit of Mass (Weight)

Unit	Abbreviation	Length (Distance)
ton	t	1 000 kilograms
kilogram	kg	1 000 grams
hectogram	hg	100 grams
decagrams	dag	10 grams
gram	g	1 gram
decigram	dg	0.1 gram
centigram	cg	0.01 gram
milligram	Mg	0.001 gram

Unit of Capacity (Volume)

Unit	Abbreviation	Length (Distance)
kiloliter	kL	1 000 L
hectoliter	hL	100 L
Decaliter	daL	10 L
liter	L	1 L
deciliter	dL	0.1 L
centiliter	cL	0.01 L
milliliter	mL	0.001 L

[27] [29]

METRIC CONVERSIONS

Unit of length

The standard unit in measuring length is a meter. We use many units to determine lengths. The kilometer is a greater unit of length measurement. Smaller measurement units include the decimeter, centimeter, and millimeter.

1 m =	10 dm =	100 cm =	1000 mm
0.1 m =	1 dm =	10 cm =	100 mm
0.01 m =	0.1 dm =	1 cm =	10 mm
0.001 m =	0.01 dm =	0.1 cm =	1 mm

1 km =	10 hm =	1000 m =	
0.1 km =	1 hm =	100 m =	
0.01 km =	0.1 hm =	10 m =	
0.001 km =	0.01 hm =	1 m =	

UNIT OF SURFACE

36

The surface area is the total exposed area of a given boundary. The standard unit of area is square meter (m^2).

Surface Area of a Square

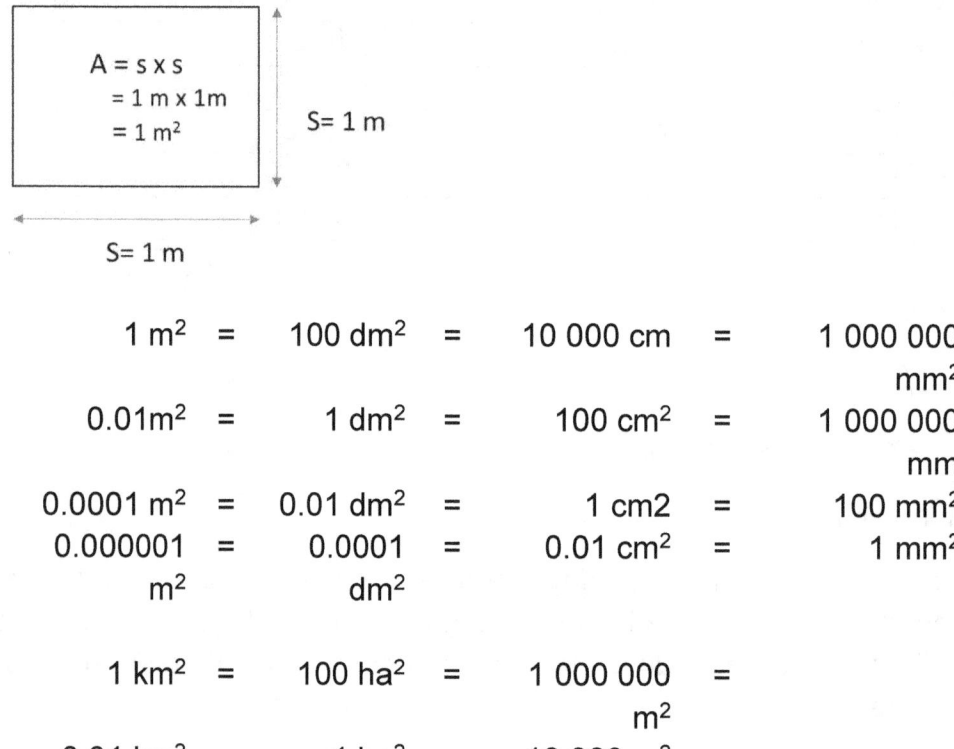

1 m^2	=	100 dm^2	=	10 000 cm	=	1 000 000 mm^2
0.01 m^2	=	1 dm^2	=	100 cm^2	=	1 000 000 mm
0.0001 m^2	=	0.01 dm^2	=	1 $cm2$	=	100 mm^2
0.000001 m^2	=	0.0001 dm^2	=	0.01 cm^2	=	1 mm^2
1 km^2	=	100 ha^2	=	1 000 000 m^2	=	
0.01 km^2	=	1 ha^2	=	10 000 m^2	=	
0.000001 km^2	=	0.0001 ha^2	=	1 m^2	=	

Note :
1 ha = 100 m x 100 m = 10 000 m^2

[16] [22]

DECIMALS, FRACTIONS, PERCENTAGE, AND RATIO

Decimals, Fractions, and Percentages are just different ways of showing the same value.

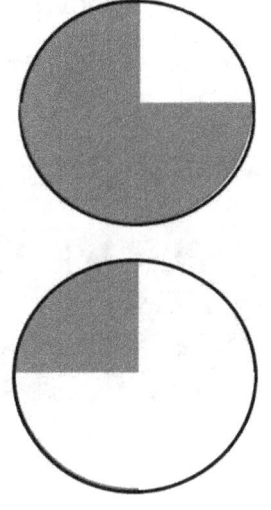

A Three quarter can be written
… as a fraction : ¾
… as decimal : .75
… as a percentage : 75%

A Three quarter can be written
… as a fraction : ¾
… as decimal : .75
… as a percentage : 75%

Example Values

Table for commonly used values shown in percent, decimal and fraction form:

Percent	Decimal	Fraction
1%	0.01	1/100
5%	0.05	1/20
10%	0.1	1/10
12.5 %	0.125	1/8
20%	0.2	1/5
25%	0.25	1/4
33.5%	0.333	1/3
50%	0.5	½
75%	0.75	3/4
80%	0.8	4/5
90%	0.9	9/10
100%	1	

CONVERSIONS

From Percent to Decimal

To convert percent to decimal, divide by 100 and remove the % sign

An easy way to divide by 100 is to move the decimal point 2 places to the left.

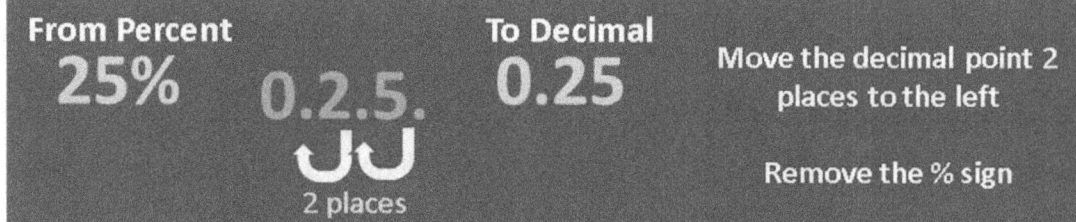

From Decimal to Percent

To convert from decimal to percent multiply by 100%

An easy way to multiply by 100 is to move the decimal point 2 places to the right:

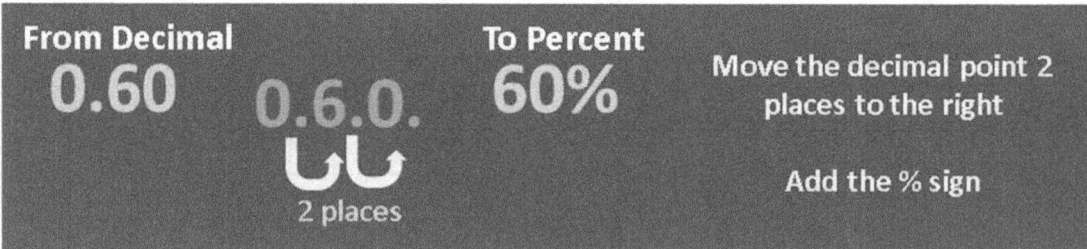

From Fraction to Decimal

To convert a fraction to a decimal divide the top number by the bottom

Example : Convert 5/8 to a decimal

Divide 5 by 8 : 5 ÷ 8 = 0.625

Answer: 5/8 = 0.625

From Decimal to Fraction

To convert a decimal to a fraction, needs to do the following

1. Write down the decimal "over" the number 1;
2. Multiply top and bottom by 10 for every number after the decimal point
   (10 for 1 number, 100 for 2 numbers, etc.)
3. Reduce the fraction. Find the greatest common factor of the numerator and denominator and divide both by the GCF
4. Simplify the fraction [33] [30] [20]

Example: Convert 0.8 to a fraction

Write down the decimal "over" over the number 1	$\dfrac{0.8}{1}$
Multiply top and bottom by 100 (2 decimal point)	$\dfrac{0.8 \times 100}{1 \times 100}$
Find the fraction, reduce.	$\dfrac{80}{100}$
Simplify	$\dfrac{4}{5}$

From Fraction to Percentage

Divide the top number by the bottom number, then multiply the result by 100 percent to convert a fraction to a percentage.

Example: Convert 7/8 to a percentage

First divide 7 by 8	$\dfrac{7}{8}$
Then multiply by 100%:	$0.875 \times 100\%$
Answer : 3/8	= 8.75

From Percentage to Fraction

To convert a percentage to a fraction, first convert to a decimal (divide by 100), then use the steps for converting decimal to fractions

Example: Convert 75% to a fraction

Convert 75% to a decimal (= 75/100)	= 0.75
Write down the decimal "over" the number 1:	$\dfrac{0.75}{1}$
Multiply top and bottom by 10 for every number after the decimal point (10 for 1 number, 100 for 2 numbers, etc.)	$\dfrac{0.75 \times 100}{1 \times 100}$
Fraction formed	$\dfrac{75}{100}$ or $\dfrac{75}{100}$

[23] [34] [33]

Chapter 3
Interpreting Plans and Drawings

Learning Outcomes

1. Layout garden plots

    1.1. Design farm plans and layouts according to crop grown

    1.2. Follow strictly planting system and practices according to approved cultural practices

    1.3. Interpret irrigation system plan according to established procedures

    1.4. Differentiate designs of irrigation systems

Farm Plans and Layout

Farm plans are crucial since they give a roadmap for managing the farm's operations and ensure that everything runs smoothly. Its goal is to allocate agricultural resources to plan production so that the farmer's resource efficiency and entrepreneurship are increased.

The size, shape, and quantity of fields and other farm facilities must all be considered while creating a farm layout. When designing or reorganizing a farm layout, the most important considerations are convenience and cost of operation. An ideal farm layout is set up for routine agricultural work so that there is minimal time wasted, no retracing of steps, and no lost motion. [1]

Considerations for a Farm Plan

1. Site Assessment
   An on-site examination of a farm is required to create a map based on the terrain, boundaries, soil, water resources, and other property features. A site evaluation will also aid in the creation of a farm business plan.

2. Site Selection
   It's crucial to find the right location. For optimal sunlight, warmth, and wind shelter, slopes to the northeast are preferable. Slopes are prone to erosion and must be managed carefully. Vegetables should not be grown on slopes more than seven degrees (7°) to avoid soil erosion and river silting. [2] [3]

3. Buffer Zone
   Buffer zones are vegetated areas that must be constructed or left to safeguard sensitive environmental areas and give wildlife habitat. They should be placed between the farm's activity and any regions where water quality may be harmed or contaminated.
   - Riparian areas include flood plains adjacent to rivers and streams and other watercourses.
   - Wetlands (areas of land covered in shallow water, either temporarily or permanently, and play an important role in nutrient recycling) Wetlands are high in ecological productivity and should not be drained, filled, or used as storage areas.
   - Drainage lines. Allow at least 20 meters of cleared natural vegetation between the top of the bank and your farming activities for small streams and 50–100 meters for rivers. Farm chemical and nutrient runoff will be intercepted and filtered in this buffer zone. [3] [4]
   -

4. Soil Type
   The type of soil used should be appropriate for the crop being cultivated. Deep, well-drained sands, sandy loams, and loams are the best soil types. Heavy clays are less suited since they have poor

drainage and are prone to waterlogging. If in doubt, a physical examination of the soil type should be performed.

5. Groundwater Contamination
   Check for contamination in the groundwater. Pollution of the groundwater might have a negative impact on your farm's operations. Although well-drained soils are ideal for growing vegetables, pesticides, herbicides, and fertilizers are more likely to leak through and damage groundwater supplies.

6. Windbreaks
   Wind protection and screening are advised for the entire farm as well as specific large paddocks. Many objections regarding farm activities are avoided by screening the property, especially with trees. Windbreaks also keep sprays, dust, bugs, and noise from adjoining areas.

7. Soil Management
   Your soil management techniques must be planned. It's critical to create grassed drainage/waterways in conjunction with fields built to avoid erosion from irrigation and heavy rain.

8. Water Management
   Pre-cropping assessment is required for water management techniques. During times of drought, when water demand is highest, water sources must be sufficient to meet the needs of crops. Poorly planned water supplies will limit crop yield and profitability. [5] [6]

SYSTEM PLANTING

The system planting method is suitable for planting fruits in all types of weather. The following are some of the most prevalent planting systems for fruit trees:

1. Square System
The trees in the square system are planted at the four corners of a square, with the same space between rows and between plants

in the same row. This is the most basic and straightforward system of the plantation.

Advantages
1. Straightening irrigation canals and routes is possible.
2. Plowing, harrowing, cultivating, spraying, and harvesting becomes simple operations.
3. Better orchard supervision is possible because one can see the orchard from one end to another.

Disadvantages
1. There will be a reduction in the number of trees that can be accommodated.
2. Because the distance between plants and rows remains constant, a portion of the space in the center of four trees is wasted. [7]

2. **Rectangular System**
The trees in a rectangle system are planted similarly to those in a square system, except that the spacing between rows is greater than the gap between plants in the same row. In this method, four nearby trees form a rectangle pattern.

Advantages
1. Intercultural operations are simple to carry out.
2. Irrigation channels can be built in various lengths and widths.
3. The huge inter gaps between rows allow light to enter the orchard.
4. Better supervision is possible

Disadvantages
1. If intercropping is not used, a considerable portion of the orchard between rows is squandered.
2. There are fewer trees planted.

3. **Diagonal System**
A diagonal system is essentially the same as a square system, except for a tree in the center of each square. The number of

trees planted in the same area is nearly doubled with this system. However, there is a significant reduction in the spacing between the trees. As a result, shorter trees are picked for the center. The middle trees will have completed their life cycle by the time the major trees reach full growth. The center trees are referred to as filler crops, while the others are referred to as primary crops. Filler crops should be eliminated if they are substantially impeding the growth of the main trees. Filler crops in orchards with mango jack and tamarind trees include papaya, guava, lime, plum, and peaches. [8]

Advantages
1. Until the primary crop bears fruit, the filler crop can provide additional income.
2. When compared to square and rectangular systems, nearly twice as many trees can be planted at the start.
3. The land can be used to its full potential.

Disadvantages
1. Layouting the orchard requires skill.
2. Inter/filler crops can obstruct the primary crop's growth.
3. Cross-cultural operations become more challenging.
4. If the filler crop is permitted to continue growing after the main crop, the main crop spacing is lowered. [5][9]

4. **Hexagonal System**
Trees are planted in the corners of an equilateral triangle in the hexagonal arrangement. A hexagon is formed by joining six similar triangles together. Six trees are at each of the hexagon's corners, with a seventh in the center, all organized in three rows. The distance between trees in six directions from the central tree, however, remains constant.

Advantages
1. In comparison to the square method, 15% more trees can be planted.
2. It is an excellent system for land that is fertile and well watered.
3. The space between plants can be kept constant.

4. It is possible to earn more money.

Disadvantages
1. Intercultural operations are becoming more challenging.
2. The planning of the orchard requires skill. [4][10]

5. **Contour System**

    A contour is an imaginary line that runs over a slope, connecting all points of equal elevation. Many depressions, ridges, furrows, and place surfaces can be found in a hilly environment. When planting, however, a line is drawn across the slope by connecting all places of the same height from a baseline. This row maintains the same spacing. However, because the degree of slope varies from spot to spot, row to row distance will vary. With the use of a Dumpy level or another suitable instrument, points of equal elevation at a length and spacing marked from comparable plant to plant equal to plant spacing are marked.

    The rows in these systems are defined by the contour lines designated at the row to row distance. However, it is impossible to keep the row to row distance constant across the rows. When the distance between adjacent contour lines is nearly doubled, a new contour is inserted into the gap.[3] [11]

    Advantages
    1. This technique may be used in both steep and flat terrain.
    2. The contour method can help to prevent soil erosion.
    3. It contributes to water conservation at the same time.
    4. Plant nutrients given by manures and fertilizers can be preserved.
    5. Contours from easy path movements on the hill slopes are utilized for weeding, manuring, pruning, harvesting, disease and insect control, and other orchard operations.

    Disadvantages
    1. Laying down contour lines is time-consuming and complicated.

2. The structure of this system necessitates a high level of expertise.
3. Making contour lines necessitates the use of specialized instruments.
4. Because the row to row distance will not be equal, modifications to the plant to plat distance may be necessary.
5. Rows are dismantled into small pieces. [2][12]

FARM LAYOUT

A few common processes can be found in the layout methods for many systems.
1. Take measurements of the land.
2. Select the tree species to be planted. The planting distance and plating system.
3. Draw up a plan on paper, filling in all the specifics.

Preparing a plan on paper is time-consuming and challenging, but the actual layout becomes easier.
At the start of the layout, a baseline (parallel to either side of the plot or a contour line) is always marketed, establishing a row of trees. Except in the contour system, subsequent rows are marked parallel to this baseline. The wooden pegs are used to indicate the position of the trees in each row, allowing space on either side equal to half the plant to plant distance (boundary). Otherwise, the tree's roots and canopy may extend beyond the limit. Boundary space can be changed based on the length and width of the land and the interplant and interrow distance. All of these aspects are taken into account while creating the blueprint. [1][13]

GOVERNMENT PLANS

Acquaint yourself with the agricultural sector's Development Plans and Policies, as they may have short and long-term implications for your proposed or existing farm operation. This will assist you in reducing unexpected hazards and improving your agricultural business.

LAYOUT PLAN FOR IRRIGATION SYSTEM

Irrigation management, or irrigation generally, is concerned with promoting optimal plant growth and maintaining proper soil moisture levels. Another goal is to ensure that there is backup insurance in case of a short-term drought so that the field can be sustained when water levels are low. Another purpose is to keep the atmosphere and soil cool, which is suitable for plants. [13][14]

Functions of Farm Irrigation System

The primary purpose of farm irrigation systems is to provide crops with adequate irrigation water at the right time. The following are some of the specific functions:
1. Diverting water from the water source.
2. Disseminating it to particular farm fields.
3. Distributing it throughout several fields.
4. Providing a system for measuring and controlling flow rates.

Reasons for An Irrigation Plan

- The designer can lay out the irrigation system in the most cost-effective method using a project plan. The plan is used to create a material list and assess the project's expected expenditures.
- The plan includes detailed instructions for setting up the system. The plan should include information on crop spacing, sprinklers, pumping requirements, and pipeline sizes and lengths. Roads, trees, gas, oil, water, telephone, and transmission lines are important barriers to note.
- Specifications, design standards, work timelines, and any contractual agreements between the installation contractor and the farmer are outlined in a plan. The plan serves as a record for future use.

- It can be used to plan a farm's overall operation and determine the limits of expansion potential.

Types of Irrigation System

Irrigation systems come in various shapes and sizes, depending on how the water is distributed around the area. Irrigation systems come in a variety of shapes and sizes. [12] [15]

TYPES OF IRRIGATION SYSTEMS

Surface Irrigation
Water is distributed over and across land by gravity, without the use of a mechanical pump.

source: aboutcivil.org

Localized Irrigation
Water is distributed under low pressure and applied to each plant via a piped network.

source: prakor.com

Drip Irrigation
Drops of water are given at or near the root of plants in this sort of targeted irrigation. Evaporation and runoff are reduced in this method of irrigation.

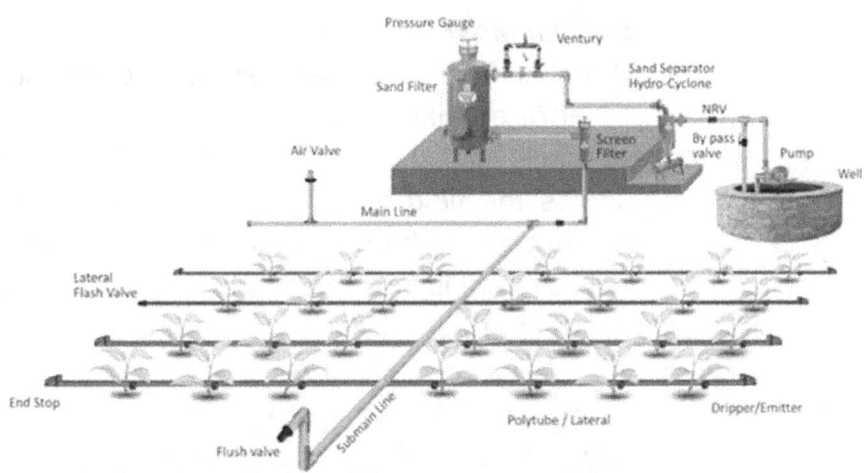
source: plantationsinternational.com

Sprinkler Irrigation
Water is distributed from a central position in the field by overhead high-pressure sprinklers or guns, or from sprinklers on movable platforms.

Several sprinkler heads are connected to a lateral pipe, which is supplied by a mainline. Source: ACCESS IRRIGATION (n.y.)

Center Pivot Irrigation
A system of sprinklers that travel in a circular pattern atop wheeled towers distributes the water. In flat areas of the United States, this approach is prevalent.

Lateral Move Irrigation
Water is supplied by several pipes, each with a wheel and a set of sprinklers that can be moved manually or by a specially designed mechanism. The sprinklers move across the field for a certain distance before the water hose needs to be reconnected for the following distance. This system is less expensive than others, but it involves more labor.

source: usgs.gov

Sub-irrigation

Water is spread across land by pumping stations, canals, gates, and ditches that raise the water table. In places with high water tables, this method of irrigation is most effective.

source: insideurbangreen.typepad.com

Manual Irrigation

Manual labor and watering cans are used to spread water across the area. This system necessitates a lot of effort.[11][16] [

source: nobowa.com

ESSENTIAL FEATURES OF A PLAN

- Topographic Data - The field shape must be precisely drawn, including all relevant obstructions, features, and elevation characteristics.

- Water Source Capacity - The location and capacity of the water source must be clearly documented.
- A well log or water license, depending on the water source, must be included with the irrigation plan. Water Management Branch licensing is also required for irrigation reservoirs.
- Soil and Crop Constraints - In order to prevent runoff and deep percolation caused by irrigation system mismanagement, soil and crop limitations must be taken into account.
- Design parameters - To choose the right irrigation system design, consider soil water holding capacity, maximum application rate, and climate data.
- Design Data - The plan must show the nozzle selected, operating pressure, discharge rate, and sprinkler spacing. Calculate the irrigation interval, set time, application rate, and net quantity applied.

Chapter 4
Safety measures in Farm Operations

Learning Objectives

1. Apply appropriate safety measures while working on the farm

    1.1. Apply safety measures based on work requirement and farm procedures

    1.2. Utilize tools and materials following farm requirements

    1.3. Explain the importance of wearing work outfit per farm requirements

    1.4. Check shelf life and/or expiration of materials and supplies against manufacturer's specifications

    1.5. Identify risks and hazards in the workplace and report them per farm guidelines

2. Safely keep/dispose of tools, materials, and outfit

    2.1. Observe procedures in cleaning used outfits in line with farm procedure before storing

    2.2. Follow the guidelines in unused labeling materials to be stored according to the manufacturer's recommendation and farm requirements.

    2.3. Dispose of waste materials according to manufacturer's, government, and farm requirements [1] [10]

APPLYING SAFETY OPERATION

Definition of Terms

Safety - deals with the physical or environmental work conditions that meet the specified Occupational Health and Safety (OHS) standards and allow workers to perform their duties without or within acceptable levels of risk.

Occupational safety - refers to the methods used in the manufacturing and labor processes.

Health is a sound state of the worker's body and mind that usually allows them to do their job.

Sharpening - is the process of thinning the edge of knives, pruning shears, and other similar items.

Cleaning - is the act or process of eliminating dirt from farm equipment, containers, and tools.

Disinfection chemicals - are chemicals that are used in cleaning and can kill bacteria. [2] [9]

Apply appropriate safety measures while working on the farm

1.1. Apply safety measures based on work requirement and farm procedures
1.2. Utilize tools and materials in accordance with farm requirements
1.3. Explain the importance of wearing work outfit in accordance with farm requirements
1.4. Check shelf life and/or expiration of materials and supplies against manufacturer's specifications
1.5. Identify risks and hazards in the workplace and report them per farm guidelines [3]

Hazard, Risk And Exposure In The Farm

Agricultural crop production entails a variety of tasks that must be completed at various locations. We put ourselves in a lot of danger while engaging in these activities. A major cause of accident, injury, or harm to a person who performs such a duty is a workplace hazard. These dangers should be a top priority for everyone involved in a particular activity or task. When it comes to risk management, it's critical to understand the differences between hazard, risk, and exposure. [4]

- A hazard is defined as the possibility of injury or a negative impact on an employee's health.
- Danger is something that could cause injury or illness to anyone at or near a workplace.
- Risk refers to the possibility that danger will cause injury or illness to anyone at or near a workplace. The severity of the hazard and the duration and frequency of exposure enhance the amount of risk.
- Exposure occurs when a person comes into contact with a hazard

Classes of Hazard

Hazards can be categorized into five different categories. They are as follows:

1. Physical hazards - includes floors, stairs, work platforms, steps, ladders, fire, falling objects, slippery surfaces, manual handling (lifting, pushing, pulling), excessively loud and prolonged noise, vibration, heat and cold, radiation, poor lighting, ventilation, air quality

2. Mechanical and electrical hazards - includes electricity, machinery, equipment, pressure vessels, dangerous goods, forklifts, cranes, hoists

3. Chemical hazards - includes chemical substances such as acids or poisons and those that could lead to fire or explosion, like pesticides, herbicides, cleaning agents, dusts, and fumes from various processes such as welding

4. Biological hazards - includes bacteria, viruses, mold, mildew, insects, vermin, animals

5. Psychosocial environment hazards - include workplace stressors arising from a variety of sources. [5] [6]

Farm emergency procedures regarding safety working environment

1. Identify the potential emergencies. The emergencies that may occur on a crop production farm could include:

    a. fire
    b. Flood
    c. typhoon,
    d. machinery entrapment
    e. electrical shock,
    f. snake or spider bite
    g. chemical exposure,
    h. injuries,
    i. illness and
    j. accidents.

2. Provide suitable emergency facilities for the types of crises that may occur on the farm (e.g., deluge showers, eyewashes, firefighting equipment, and first aid kits).
3. Ensure you have the right equipment on hand to contain and clean up any chemical or other hazardous substance spills that may occur.
4. People working on and visiting the farm should be aware of and understand the emergency protocols and their obligations to reduce the risk of personal harm or property damage in the event of an emergency.
5. Teach everyone who works on the farm how to respond to an emergency.
6. Everyone should be aware of where fire alarms, fire extinguishers, and first-aid kits are located, as well as how and where to contact emergency services and where to safely congregate in the event of an emergency. [6]

The following factors may increase the risk of injury or illness for farmworkers:

1. Age – Injury rates are highest among children age 15 and under and adults over 65.
2. Equipment and Machinery – Most farm accidents and fatalities involve machinery. Proper machine guarding and doing equipment maintenance according to manufacturers' recommendations can help prevent accidents.

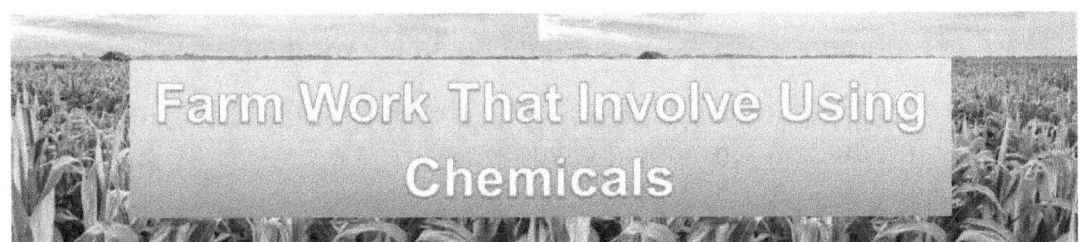

1. Spraying Chemicals
   Pesticides are just one of the many chemicals utilized on a farm. These compounds are used to nourish and control pests like insects, weeds, and mollusks, among other things. The majority of these chemicals are sprayed on.

   Examples of chemical hazards:
   A. Spraying in a high wind causes the spray to float over a dam or the farmhouse.
   B. Water from spraying equipment running into open drains, puddles, or stockyards or dams.
   C. Chemicals or containers that have been left lying around. Empty canisters piled on top of one other.

   Chemical risks can be reduced in a variety of methods, including:
   A. Put on safety gear like respirators, waterproof clothing, rubber gloves, and waterproof shoes.
   B. Ensure that chemicals are stored safely and that cupboards are locked.
   C. Never spray chemicals if there is a strong wind.

D. Be familiar with first-aid procedures.
E. Keep track of all hazardous chemicals used on the farm. [7]

Chemicals should be used safely.
A. Determine whether a chemical substance is genuinely required.
B. Remove a dangerous substance or replace it with one that is less hazardous if it isn't practicable.
C. Use of safe work practices or personal protection equipment.
D. Keep track of the chemicals used on the farm.

PERSONAL PROTECTIVE EQUIPMENT (PPE)

Personal protection equipment (PPE) can help to limit the number and severity of accidents and illnesses associated with farm work. Personal protection equipment helps keep people safe, and it also helps increase productivity and revenues. When the task and its possible hazards require it, farmers and ranchers can share in these benefits by wearing the proper protective equipment for themselves, family members, and employees. [8]

- When working on construction sites, pruning trees, fixing machinery, and other occupations that pose a risk of head injury, wear a hard hat.
- To prevent skin cancer, use a sun safety hat (one with a wide brim and neck protection).
- When applying pesticides, fertilizers, working in the shop, or in strong dust conditions, wear proper protective eyewear (safety glasses, goggles, face shields).
- When operating noisy equipment such as grain dryers, feed grinders, older tractors, chain saws, and so on, use acoustic earmuffs or plugs to protect your hearing.
- Protect your lungs with the correct respiratory equipment (dust masks, cartridge respirators, gas masks, air packs) when working in dusty or moldy conditions, spray painting, applying chemicals, working in bins, tanks, silos, and manure storage places. [9]

Source: suresafety.com

urce: au.prochoicesafetygear.com

Source: cleansafefeltd.com

Basic First Aid

Recognizing the situation and offering assistance are the first steps in an emergency. Always activate the emergency response system by contacting your locality's Emergency Medical Services (EMS) number when in doubt or when someone is critically injured or ill.

Stay on the line until more assistance arrives or the EMS operator tells you to hang up. Dispatchers can walk you through conducting cardiopulmonary resuscitation (CPR), using an automated external defibrillator (AED), or providing basic treatment until more assistance comes.

Know where the first-aid kit and the AED are stored and what they include. Know how to call your local EMS (Emergency Medical Services). Be aware of any workplace policies about medical emergencies.

After determining the problem, the next step in offering assistance is to assess the injured or unwell person's response. The best method to tell is to tap the individual on the shoulder and ask loudly, "Are you okay?" Yell for assistance after determining responsiveness. Any medical identification, such as a necklace or a bracelet, should be looked for. This could be a crucial indicator as to what's causing the problem. [10]

FIRST AID KIT

Common items found in a first aid kit are:

- Bandages, roller bandages, and tape
- Sterile gauze
- Antiseptic wipes and swabs
- Absorbent compresses
- Antibiotic cream
- Burn ointment
- Mask for breathing (rescue breathing/CPR)

- Chemical cold pack
- Eyeshield and eyewash
- First aid reference guide that includes local phone numbers [11]

FARM EMERGENCY PLAN

Build Your Emergency Plan
1. Make an inventory of all your assets, properties, and list of hazards
   a. Take a tour around your property and make a list of all the potential hazards. Make a list of the type of hazard, the name of the facility or site, and the type of the first responder who will be dispatched to counter the threat. The location of stored chemicals, insecticides, hazardous products, and combustible materials should be meticulously documented.
   b. Make a list of all assets, including cattle, feed, tractors, and combines. Write down the serial numbers for your machinery, as well as medicines for the cattle. Now that you know what might happen and where it might happen, you can devise a tailored strategy to combat each of these threats.
2. Leverage Aerial Maps to Mark Locations
   a. Print out a satellite image or map of your property – Take notes and markings on the map so that the first responders will know exactly what they're looking. at
   b. Identify structure and Hazards – from the list of hazards you made, number all the buildings and site location
   c. Label street and road – name the access routes, gates, entry and exit locations.
   d. Crop and water = identify the types of crops growing up and water sources too.

3. Emergency Contacts and Communication
   a. Define roles and responsibilities - For each of the emergency scenarios you've defined above, assign staff a task. Review these plans frequently, and spend time at meetings going over who is involved in each scenario.
4. Relocate Hazardous and explosive materials

a. Examine the spot where you keep your volatile things. Consider how close these items are to cattle, as well as family and staff living and working areas. To lessen the threat that dangerous or explosive materials offer, it may be essential to transfer them. In addition, assign storage places on the map that are less likely to catch fire or cause damage to neighboring structures in the case of a chemical emergency.

5. Map out Evacuation Routes- Getting away safely is contingent on the circumstances. Depending on the type of disaster, livestock may need to be evacuated to different regions. If poisonous substances threaten animals or humans, moving upwind and away from an otherwise defined evacuation zone may be necessary. Weather-related incidents may necessitate alternative arrangements. [12]

CLEANING, STORING, AND WASTE MANAGEMENT

Protect Tools From the Elements

Lubricant oil can be sprayed on electric hedge trimmer blades, hoes, shovels, and other metal surfaces. First, spray the blades, then turn them on to ensure that the oil gets into all of the crevices. All electrical and petrol gardening equipment must be covered with a blanket or sheet is placed in the shed. This will keep dust and grime away from them.

The farmer and/or agricultural workers in charge of cleaning must follow the following processes as closely as possible:

- Be thoroughly versed in cleaning techniques.
- Develop a cleaning program and schedule it according to the recommended frequency, and ensure that the cleaning program is effective by monitoring it.
- Cleaning should not occur when fresh veggies are being picked, packed, handled, stored, and cleaning water should be safe.
- Cleaning and disinfection chemicals must be used in a designated area away from the field and store agricultural inputs and fresh vegetables. When utilizing cleaning and

disinfection chemicals, the farmer and/or farm employees must become familiar with the instructions for use.
- Follow all precautionary cautions and mixing instructions to the letter.
- When working with chemicals, protect your equipment, tools, containers, and fresh veggies; familiarize yourself with the product's instructions for use. [7] [11]

Cleaning reusable containers:

The farmer and/or farm employees in charge of cleaning reusable containers must follow the following processes as closely as possible:

• Remove as much plant debris, soil, and residues as possible, using a brush or appropriate equipment as needed.

• Check containers for physical damage that could harm, deteriorate, or contaminate fresh veggies, and repair them if necessary.

• Check containers for any missed plant debris, soil, or residues, and re-clean if necessary.

• If cleaning and/or disinfection chemicals are used, follow the label's mixing recommendations.

• Clean the containers by rinsing them in clean water.

• Dry them as quickly as possible by placing them in the sun.

- To avoid infection, store reusable containers correctly. [3] [10]

Cleaning Equipment, Tools And Garbage Cans

The farmer and/or farm employees in charge of cleaning reusable containers must follow the following processes as closely as possible: • Remove as much plant debris, soil, and residues as possible, using a brush or appropriate equipment as needed.

• Check containers for physical damage that could harm, deteriorate, or contaminate fresh veggies, and repair them if necessary.

• Check containers for any missed plant debris, soil, or residues, and re-clean if necessary.

• If cleaning and/or disinfection chemicals are used, follow the label's mixing recommendations. [4] [9]

Cleaning Areas for Handling and Storing Fresh Produce

The farmer and farm employees in charge of cleaning these areas must follow the following steps as closely as possible:

1. Unplug any electrical equipment and, if feasible, cover electrical motors, electrical boxes, connections, light fixtures, and other electrical equipment with plastic. This task should not be completed with packaging materials.
2. Eliminate garbage and any accumulated plant debris from the floors, and rinse the entire ceiling and light fixtures with low-pressure water to remove dust and soil buildup. Wash the walls, windows, and doors from top to bottom. Remove any soil buildup by rinsing the whole floor surface. Make sure you don't get any water on your equipment.
3. Scrub regions with a brush and cleaning chemicals such as detergent if necessary, making sure no spots are missed.

4. Wash out drains after washing regions with cleaning products; take care not to spray water on equipment. If cleaning and/or disinfection chemicals are used, follow the label's mixing recommendations.
5. Clean the containers by rinsing them in clean water.
6. Dry them as quickly as possible by placing them in the sun.
7. To avoid infection, store reusable containers correctly. [8] [9]

Cleaning hygienic facilities:

The farmer and/or farm workers responsible for cleaning hygienic facilities must adhere as much as possible to the following procedures:

- Pick up trash from the floors and in a trash can.

- By using proper detergents, clean toilets, sinks, and any other fixtures.

- Using low-pressure water, rinse the entire floor surface to remove any soil build-up.

- If cleaning and/or disinfection chemicals are used, follow label instructions for mixing.

- As required, apply cleaning materials or disinfection chemicals to the entire floor surface area, scrub areas with a brush if needed, and ensure that no spots are missed.

- Rinse floor and drains.

- Remove excess water and allow drying out at room temperature.

- Ensure that hygienic facilities have enough toilet paper, soap, and disposable towels. [2] [7]

Technique in storing chemicals

Farms employ chemicals for a variety of reasons. Access to knowledge and responsible action are required for chemical management to be safe. Farm chemical producers, suppliers, and users all play a vital role. Chemical chemicals pose a variety of health, safety, and environmental dangers to individuals. As a result, they are subject to a variety of regulations. These rules are intended to ensure that chemicals are used safely and efficiently, with the least amount of risk to human health, the environment, and property damage.

Safe Management of chemicals involves:

- correct labeling and packaging;
- provision of material safety data sheets (MSDS);
- safe transport, storage, use, and disposal of substances.

Labeling and Packaging of Chemicals

Chemicals must be supplied in packages that are correctly labeled and suitable for the substance. Information provided on the label will depend on the type of substance and the risks associated with it. Items to look for are:

1. Signal words such as _CAUTION ', _POISON 'or _DANGEROUS POISON ', used for scheduled poisons – a signal word alerts users to the possibility of poisoning if the substance is swallowed, inhaled, or absorbed through the skin.
2. The Dangerous Goods (ADG) diamond, if there is an immediate risk to health or safety e.g., flammable liquids.
3. Risk phrases describing the type of health effects e.g., _irritating the skin ', and safety phrases stating precautions for safe handling, storage, spills, disposal and fire e.g., _keep away from combustible material.'[6] [10

Ensure that containers remain labeled

Farmers must make sure that original labels remain on chemical containers. If a chemical is poured into a second container, such as a spray tank, the product name, and necessary risk and safety terms must be labeled on that container. In most cases, you can copy these from the parent container. If a chemical is consumed right away and the container is well cleaned, no labeling is required.

There are several solid reasons to use suitable containers and labeling, including:

• Storing poisons in food containers can result in poisoning due to unintentional ingesting.

• If something goes wrong and an unlabeled container is to blame, insurance firms may dispute liability.
• Produce cannot be exported if maximum residual limitations are exceeded. Labels provide information on the permitted use and withholding periods for agricultural and veterinary chemicals.[11][5]

Material Safety Data Sheets

The Material Safety Data Sheet (MSDS) is more than just a piece of paper. It contains vital information regarding the product's ingredients, health effects, safe use and handling, storage, disposal, first aid, and emergency procedures. Farmers must get MSDSs from their suppliers and store them in a register accessible to everyone who may be exposed to the hazardous product.

Chemical Storage and Transportation

Farm chemicals must be stored safely to keep them safe from the elements, limit access to them, avoid contamination of the environment, food, or livestock, and keep them separate from other incompatible chemicals.

There must be plans in place to contain any chemical spills. For example, a farmer may determine that a secured shed with a roof and concrete floor, which is bounded to contain any spills, is the best approach to ensure safe storage after assessing the potential damage to people's health or the environment.

Always keep in mind that oxidizing substances should never be stored with fuels. That is, chemicals labeled with a yellow diamond and a red diamond should never be stored together.

The safe transport of farm chemicals depends on the substance, quantity, location to which it will be delivered, and what else will be transported with it. Small volumes (less than 250 liters) can generally be transported by vehicle if the container is adequately secured and spill-proof. [4] [11]

ENVIRONMENTAL LAW OF THE PHILIPPINES

The Philippine Environmental Code, promulgated by Presidential Decree (PD) 1152 in 1977, lays the groundwork for a comprehensive waste management plan covering everything from waste generation to disposal methods. In addition, PD 1152 mandates specific guidelines for the management of municipal waste (solid and liquid), sanitary landfill and incineration, and disposal sites in the Philippines.

The Toxic Substances, Hazardous and Nuclear Wastes Control Act, also known as Republic Act (RA) 6969, was enacted by the Philippine Congress in 1990 to respond to the growing problem of toxic chemicals, hazardous and nuclear wastes. The Control and Management of Toxic Substances and Hazardous and Nuclear Wastes Act (RA 6969) mandates the control and management of toxic substances, hazardous and nuclear wastes imported, manufactured, processed, distributed, used, transported, treated, and disposed of in the country.

The Act aims to safeguard public health and the environment in the Philippines from the harmful effects of these substances. Aside from RA 6969's basic policy rules and regulations, hazardous waste management must also adhere to the requirements of other specific environmental laws, such as PD 984 (Pollution Control Law), PD 1586 (Environmental Impact Assessment System Law), RA 8749 (Clean Air Act), and RA 9003 (Ecological Solid Waste Management Act), as well as their implementing rules and regulations. [3] [12]

On-farm water management

On-farm water management can be defined as: A systems approach towards controlling water on a farm in a manner that provides for the beneficial management of water for satisfying the irrigation and drainage needs of a crop under the constraints imposed by the prevailing physical social, governmental, and production systems .[13]

Water, or the control of water, affects most crop production activities. Sufficient water must be present in the root zone for germination, evapotranspiration, nutrient absorption by roots, root growth, and soil microbiological and chemical processes that aid in the decomposition of organic matter and the mineralization of nutrients. These factors are all necessary for sustaining crop growth on a particular field.[14]

At the same time, the root zone must be sufficiently dry to ensure adequate aeration and root extension. The root zone must also be dry enough to allow field access for performing cultural practice activities such as planting, cultivating, fertilization, pesticide and herbicide applications, and harvesting operations. In order to realize potential yields, water movement through the soil must exist; water movement leaches excess salts from the root zone.15]

Systems Components

Components of the water management system that must be managed on a farm can be broken down into several categories.

- Irrigation

- Drainage

- Conveyance to and from fields

- Water storage

- Release of water from the confines of the farm

- Water sources and sinks

The relationships between the components are

shown schematically [14] [15]

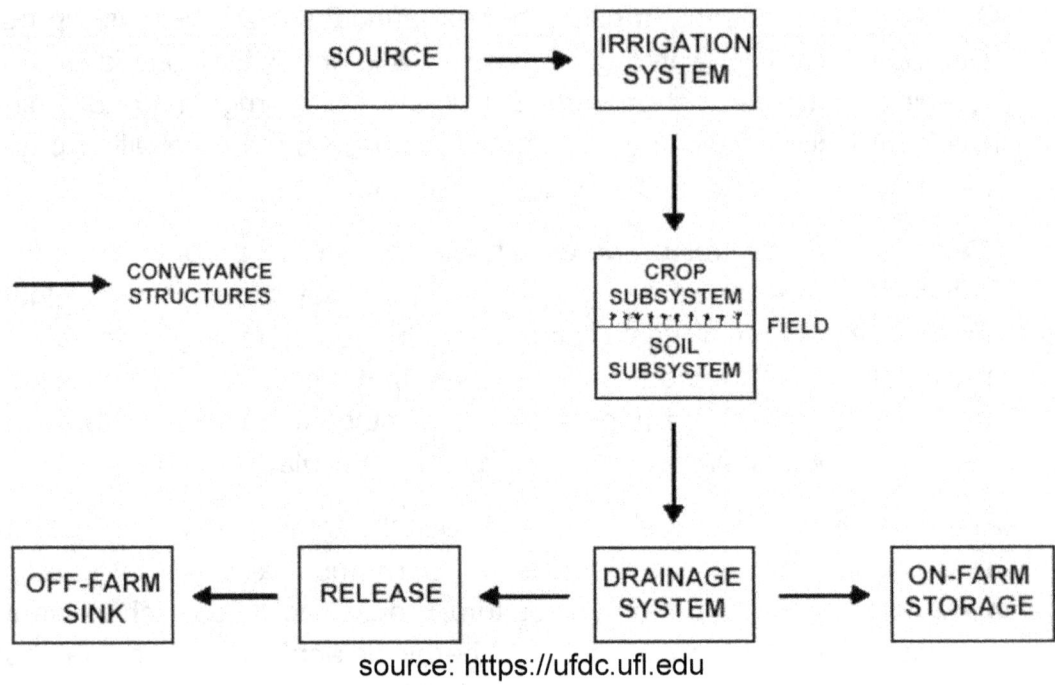

source: https://ufdc.ufl.edu

NOTES

Chapter 1

1. Agro Premio (n,d). Heavy Type Tiller Cultivator With ROller COmbination, accessed JUne 15, 2021
2. <https://agropremio.com/urun/heavy-type-tiller-cultivator-with-roller-combination/>
3. Alexbinthree (n,d). Bully Tools Spading Fork with Fiberglass D-Grip Handle- 92370 - The Home Depot, accessed June 15, 2021
4. <https://www.pinterest.ph/pin/593490057131331405/>
5. All Home (n,d). Bernmann WheelBarrow B-WBDT-PT D-TYPE, accessed June 15, 2021
6. <https://allhome.com.ph/products/bernmann-wheelbarrow-b-wbdt-pt-d-type.html>
7. Anon (n.d). Rorating Tiller Hand Tractor, accessed JUne 15, 2021,
8. <http://www.anoncn.net/product/anon-paddy-use-7hp-20hp-2wd-function-of-hand-tractor>
9. Cate's Garden (n,d). Classic Garden Hand Trowel, accessed June 15, 2021
10. <https://www.catesgarden.com/products/garden-hand-trowel>
11. Cottage Craft Works (n,d). Grape/Eye Hoes, accessed June 15, 2021
12. <https://www.pinterest.ph/pin/1348969513107372607>
13. Current Agricultue (2019). Maintenance of Simple Farm Tools, accessed June 15, 2021
14. <http://currentagric.blogspot.com/2019/11/maintenance-of-simple-farm-tools.html>
15. Daylea (n,d). G & F 10051 JustForKids Kids Water Pail with Garden Tools Set and Gloves, Gree, accessed June 15, 2021
16. <https://www.dayleasingn.com/index.php?main_page=product_info&products_id=849241>

17. Deere, J. (n,d). 5045D MFWD (4-wheel drive 45 Horse Power tractor), accessed June 16, 2021
18. <https://www.deere.com/en_ASIA/docs/brochure/product/equipment/tractors/5000_series/5d/5045d/pdf/5045d_eng.pdf>
19. ec21 (n,d). Disc Plough with 26' Plough Discs, accessed June 15, 2021.
20. <https://www.ec21.com/product-details/Disc-Plough-with-26--4034738.html>
21. European Tools Australia (n,d). Dewit midi Spade 80 cm handle, accessed June 15, 2021
22. <https://europeantoolsaustralia.com/product/midi-spade-80-cm-handle/>
23. Faithfull Quality Tools (n,d). Essentials Garden Rake, accessed June 15, 2021
24. <https://www.faithfulltools.com/p/FAIESSGRE/Essentials-Garden-Rake>
25. Farm Injury Resource Center (n,d). Equipment Design Defects, accessed JUne 15, 2021
26. <http://farminjuryresource.com/equipment-design-defects/>
27. farming-machine (n,d). Light Duty Disc Harrow, accessed June 15, 2021
28. <https://www.farming-machine.com/agricultural-machinery/tillage-implements/light-duty-disc-harrow/>
29. Filipino Bolos (n,d). Why a Bolo?, accessed June 15, 2021
30. <http://filipinobolos.com/>
31. Galleon (n,d). FITOOL Stainless Hand Fork, Stainless Steel Garden For/Weeder 3T 12-Inch-Premium Garden Hand Weeder Fork- Stainless Steel Blade Mirror Polishing- with Ash Wood Handle, accessed JUne 15, 2021
32. <https://www.galleon.ph/appliances-c863/gardening-lawn-care-c1167/pitchforks-c20523/fitool-stainless-hand-fork-stainless-steel-garden-p34705363>
33. Gardenista (n,d). Three Tine Hand Cultivator, accessed June 15, 2021
34. <https://www.gardenista.com/products/three-tine-hand-cultivator/>
35. Gransfors Bruk Sweden (n,d). Gransfors Large Carving Axe, accessed JUne 15, 2021

36. <https://www.gransforsbruk.com/en/product/gransfors-large-carving-axe/>
37. Hire Express (n,d). Mattock, accessed June 15, 2021
38. <https://www.hireexpress.com.au/product.view&pid=567>

Chapter 2

1. Antara, R (2018). Extinction of the Buffalo Tradition Plowing Rice Fields, accessed June 16, 2021
2. <https://steemit.com/photofeed/@rahmadantara/extinction-of-the-buffalo-tradition-plowing-rice-fields>
3. BCcampus (n,d). Units of Measurement, accessed June 16, 2021
4. <https://opentextbc.ca/basickitchenandfoodservicemanagement/chapter/units-of-measurement/>
5. Beaulieu, D (2020). What Do the Letters NPK Mean on a Fertilizer?, accessed June 16, 2021
6. <https://www.thespruce.com/what-does-npk-mean-for-a-fertilizer-2131094>
7. Henrietta (2020). How To Get Rid Of Weeds and Other Creeping Plants, accessed June 16, 2021
8. <https://www.tygershark.nyc/how-to-get-rid-of-weeds-and-other-creeping-plants/>
9. lumen (n,d). Units of Measurement, accessed June 16, 2021
10. <https://courses.lumenlearning.com/boundless-chemistry/chapter/units-of-measurement/>
11. Math is Fun (n,d). Decimals, Fractions and Percentages, accessed June 16, 2021
12. <https://www.mathsisfun.com/decimal-fraction-percentage.html>
13. Ministry of Agriculture, Animal Industry and Fisheries (2020). Over 1000 farmers signed up for Micro Scale Irrigation Program, accessed June 17, 2021
14. <https://www.agriculture.go.ug/over-1000-farmers-signed-up-for-micro-scale-irrigation-program/>
15. Nelson, W (2018). Top 5 Compact Tractor Attachements for Planting, accessed June 16, 2021
16. <https://nelsontractorco.com/attachments-for-planting/ >

17. Rice Knowledge Bank (n,d). How To Manage Soil Fertility, accessed June 16, 2021
18. <http://www.knowledgebank.irri.org/step-by-step-production/growth/soil-fertility>
19. Rhoades, H (n,d). When and How to Transplant Seedlings Into The Garden, accessed June 16, 2021
20. <https://www.gardeningknowhow.com/garden-how-to/propagation/seeds/when-to-transplant-a-seedling-plant-into-the-garden.htm>
21. Spengler, T (n,d). How to Save Seedlings- Troubleshooting Common Seedling Issues, accessed June 16, 2021
22. <https://www.gardeningknowhow.com/garden-how-to/propagation/seeds/troubleshooting-seedling-issues.htm>
23. SVZ Emerging Uniqueness (2021). A Farmer's perspective : The potential benefits of biodegrable mulching paper, accessed June 17, 2021
24. <http://www.svz.com/a-farmers-perspective-the-potential-benefits-of-biodegradable-mulching-paper/>
25. Syngenta (n,d). Insecticides, accessed June 16, 2021
26. <https://www.syngenta.co.in/insecticides>
27. T.L.E Learning Module (n,d). Lesson 2: Perform Estimation and Basic Calculation, accessed June 21, 2021
28. <https://gltnhs-tle.weebly.com/lesson-23.html>
29. UNDP (2013). Cambodia: Demining transforms former battleground into field of hope, accessed June 16, 2021
30. <https://reliefweb.int/report/cambodia/cambodia-demining-transforms-former-battleground-field-hope>
31. WorldAtlas (n,d). Top Pesticide Using Countries, accessed June 17, 2021
32. <https://www.worldatlas.com/articles/top-pesticide-consuming-countries-of-the-world.html>
33. Wylie House News and Notes (2017). The History of Seed Saving, accessed June 16, 2021
34. <https://libraries.indiana.edu/history-seed-saving>

Chapter 3

1. Centers for Diseas Control and Prevention (n,d). TYpes of Agricultural Water Use: Irrigation vs RAin-Fed Agriculture, accessed June 20, 2021
2. <https://www.cdc.gov/healthywater/other/agricultural/types.html>
3. Jamal, H. (25 March 2017). Surface Irrigation Methods - Adavnaatges and Disadvantages, accessed June 20, 2021
4. <https://www.aboutcivil.org/surface-irrigation-methods.html>
5. Kankam. T (19 May 2021). 5 Types of Irrigation, Why they are Critical to the Success of Your CRops, accessed June 21, 2021
6. <https://nobowa.com/types-of-irrigation/>
7. Plantations International (n,d). Drip Irrigation, accessed June 20, 2021
8. <https://www.plantationsinternational.com/drip-irrigation-systems/drip-irrigation/#iLightbox[postimages]/0.
9. Prakor (22 February 2019). Components Of A Localized Irrigation System (Part 1), accessed June 20, 2021
10. <https://prakor.com/en/components-of-a-localized-irrigation-system-part-i/>
11. Sphuler, D , Stauffer, Beat (n,d). Sprinkler Irrigation, accessed June 20, 2021
12. <https://sswm.info/sswm-university-course/module-4-sustainable-water-supply/further-resources-water-use/sprinkler-irrigation>
13. T.L.E Learning Module (n,d). Lesson 3: Interpret Plans and Drawing, accessed June 21, 2021
14. <https://gltnhs-tle.weebly.com/lesson-32.html>
15. USGS , Center Pivot Irrigation System in Arizona, USA, accessed June 20, 2021
16. <https://www.usgs.gov/media/images/center-pivot-irrigation-system-arizona-usa>

Chapter 4

1. American Family Insurance (n,d). Farm Emergency Management Plan Tips, accessed June 22, 2021

2. <https://www.amfam.com/resources/articles/on-the-farm/farm-emergency-management-plan>
3. CleanSafe Disposables (n,d). 3M 4277 Dust Mask. Online Images.
4. <https://www.cleansafeltd.com/shop/3m-4277-dust-mask/>, accessed June 21, 2021
5. NHCPS (n,d). First Aid, accessed June 21, 2021
6. <https://nhcps.com/lesson/cpr-first-aid-first-aid-basics/>
7. Pro Choice Safety Gear (n,d). Earmuffs. Online Images.
8. <https://au.prochoicesafetygear.com/hearing-protection/earmuffs>, accessed June 21, 2021
9. Sure Safety (n,d). Chemical Splash Safety Goggles . Online Images
10. <https://suresafety.com/chemical-splash-safety-googles.html>, accessed June 21, 2021
11. T.L.E. Learning Module (n,d). Lesson 4: Occupation Safety and Health, accessed June 21, 2021
12. <https://gltnhs-tle.weebly.com/lesson-46.html>

Bibliogragy

1. Holland ,T (2021). The 11 Best Watering Tools of 2021, accessed June 15, 2021
2. <https://www.thespruce.com/best-watering-tools-to-buy-4110233>
3. Indiamart (n,d). Heavy Duty Crowbar Tool, accessed June 15, 2021
4. <https://www.indiamart.com/proddetail/heavy-duty-crowbar-tool-20597638430.html>
5. Longree (n,d). Medium Duty 36hp, 37hp, 38hp, 3 Point Hitch Rotovator, Farm Rotovator, acessed JUne 15, 2021
6. <http://www.chinarotavator.com/rotary-tiller/medium-duty-rotary-tiller/medium-duty-36hp-37hp-38-hp-3-point-hitch.html>
7. Noble-Adams (n,d). Marolex Master 2L Pump Up Hand Sprayer, accessed June 15, 2021
8. <https://www.nobleadams.co.nz/shop/div+idyanSprayersdiv/Hand+Operated+Sprayers+%26+Accessories/Marolex+Master+2L+Hand+Sprayer.html>
9. Noble Research Institute (2017). Improve Safety on the Farm with Preventive Mantenace, accessed June 16, 2021

10. <https://www.noble.org/news/publications/ag-news-and-views/2017/january/improve-safety-with-preventative-maintenance/>
11. PennState Extension (2016). Pre-Operational Checks for Tractors, accessed June 16, 2021
12. <https://extension.psu.edu/pre-operational-checks-for-tractors>
13. Tabor Tools (n,d). Tabor Tools S3A Classic Pruning Shears, accesssed June 15, 2021
14. <https://store.tabortools.com/products/s3-classic-pruning-shears>
15. The City Edition (n,d). Beginner's Guide to Gardening & Smallscale Farming, accessed June 15, 2021
16. <https://www.thecityedition.com/2012/Farming_Page4.html>
17. The Toolhouse Inc (2017). Preventive Maintenance for Your Power Tools, accessed June 16, 2021
18. <https://toolhse.com/preventive-maintenance-for-your-power-tools/>
19. The Werks C & C Inc (2019). 5 Preventive Maintenance Tips for Machine Tools, accesssed June 16, 2021
20. <https://thewerkscandc.com/5-preventative-maintenance-tips-for-machine-tools/>
21. T.L.E Learning Module (n,d). Lesson 1: Use Farm Tools and Equipment, accessed June 21, 2021
22. <https://gltnhs-tle.weebly.com/lesson-14.html>
23. Tractor Supply Co (n,d). Mini Shovel Fiberglass Handle, accessed June 15, 2021
24. <https://www.tractorsupply.com/tsc/product/groundwork-mini-shovel-with-fiberglass-handle>
25. Tractors & Tractor Accessories, 2020. John Deere 7R 350 Tractor, accessed June 17, 2021,
26. <https://www.no-tillfarmer.com/articles/9804-john-deere-7r-350-tractor>
27. Wiki (2019). Seed Drill, accessed JUne 15, 2021,
28. <https://simple.wikipedia.org/wiki/Seed_drill>
29. Zozhi (n,d). Centrifugal Domestic Water Pumps DTM-18 Big Capacity Flow Up to 500 L/min, accessed June 15, 2021
30. <https://www.electricmotorwaterpump.com/quality-8661461-centrifugal-domestic-water-pumps-dtm-18-big-capacity-flow-up-to-500-l-min>

www.ingramcontent.com/pod-product-compliance
Lightning Source LLC
Chambersburg PA
CBHW081012170526
45158CB00010B/3017